MARVELOUS PURSUITS

by the same author

Berta Broadfoot and Pepin the Short: A Merovingian Romance
Cautionary Tales
The Stones Remember (co-editor)
This Terible Thing Called Love (in Hebrew translastion)

MARVELOUS PURSUITS

Barbara Goldberg

Snake Nation Press
Valdosta, Georgia
1995

Acknowledgements

Poems in this book have appeared in the following publications: *American Literary Review, Colorado Review, Confrontation, Davar, Harvard Review, Light, New England Review, Poetry, Southern California Anthology, Tel Aviv Review.*

A different form of "What Got Left Behind" received the PEN Syndicated Fiction Project Award.

My gratitude to the National Endowment for the Arts, the Maryland State Arts Council, Yaddo, and the MacDowell Colony for providing me with the time and space to write many of these poems. Thanks also to Ann Darr, Laura Fargas, Elaine Magarrell, and Catherine Harnett Shaw,—good poets, good friends.

Published by Snake Nation Press
110 #2 West Force Street
Valdosta, Georgia 31601

Editors: Roberta George and Nancy Phillips

Sponsored by: The Georgia Council for the Arts
and The Porter Flemming Foundation

L. O. C. 95-67136
ISBN 09638364-4-7

For Moshe

Contents

Foreword

David Kirby

I read a lot of poetry, so the first thing I look for is something different: not night-and-day different, but different in the sense that the poet has mastered the tradition and tried to go it one better. *Marvelous Pursuits* does this with its core sound of clipped, staccato statements that occasionally bloom into full-blown dithyrambs yet always return to the terse, the laconic, the economical. At a time when the breezy, breathy poem is the norm, it is refreshing to see a poet get so much out of containment and control.

But the main thing I look for in poetry is rich satisfaction, which means the poems must always keep something secret from the reader. Georges Braque once said that the only thing that counts in art is the thing you can't explain, and certainly all great poetry is written with one foot in the sunlit meadow of the conscious mind and the other in the shadowy forest of the unconscious.

If that means taking risks, then so be it. There is a poem in this collection called "Girl Trades Baby for Prom Dress"; anyone who has never wanted something so badly that they would give up their child for it doesn't have any business reading poetry.

Certainly not poetry as marvelous, as worth pursuing, as *Marvelous Pursuits.*

David Kirby is the W. Guy McKenzie Professor of English at Florida State University. He is the author or co-author of sixteen books, including *Saving the Young Men of Vienna* (University of Wisconsin Press), *Mark Strand and the Poet's Place in Contemporary Culture* (University of Missouri Press), and *Writing Poetry* (The Writer, Inc.).

I

IT NEEDS BRAINS

Maupassant, prodigious lover, swore
he could come twenty times in a row
and never tire. He took a dancer
from the Folies Bergére six times
in an hour (there was a bookkeeper
who counted) then strode across the hall
and came thrice more with a prostitute.
He valued carnal success above his art:
"It needs brains to give another pleasure."

Same Maupassant fell for Giséle d'Estoc,
her close-cropped hair, her manly attire.
She hankered for a trapeze artiste
from le Cirque Médrane, then stabbed her
in the thigh. Giséle shared with Guy
her slimhipped women, hashish, ether.

But the only woman he ever loved was
"L'Inconnue." She haunted his imagination:
"intensely sensuous, self-controlled,
soulful, yet a coquette." What woman wouldn't cry
"C'est moi!", relieved to be recognized.

Oh Guy, come feast on me, my ambiguities
are singular. Austerity and razzledazzle.
Don't be seduced by the sheer extravagance
of numbers. You called pursuit of The Unknown
life's grand adventure. There is no other.

IN OUR APRIL BOLD

We hold on to nothing and nothing
holds us back. We simply shed

our old identities, the vocabulary
of refinement. Dear one, speak to me

in your guttural tongue, confabulate
a camel's tale. Be sure to leave

a small morsel for the gods, so they
will not behead us, just enough

to whet their chops. It seems
we spend our lives elaborately

avoiding what we like best. To hell
with that. Let's proceed with much ado.

BALLAD OF THE ID

I am your rose hips and bunting and bootleg
and I am your black bangs those devil's
spikes and I am your spine when you lie
down in ticklegrass the zipper the cock
all glistening and white and I am the stars
that roll in their sockets that blink on
the night I could be Gibraltar or jukebox
or drumstick without me you're puny and
pallid and prudish and with you are banquet
a wishbone with tassels that time in the bistro
I bit your sister and squealed on the porter
and who pulled you out of the muck of intention
who sang hosannas for spunk and risk who
shorted the sheets who stole your tom-tom
just ask and I'll tell you I did I did

TECTONICS

Once I have committed
your body to memory, I can begin
to go beneath the surface of skin,
follicle, pore. The orphanned mother
who birthed you now is mine,
that scrappy boy at loggerheads
with the world, I claim him,
and as for you, you absorb
the brunt of my father's rage.
This is what is meant to cleave,
to take on sorrows we do not own
and make them ours, so intimate
a coupling. And that is why
in love's afterglow I feel close
to tears—something has been split
open, and if the body is like
the earth's crust, the deeper
the fault, the more it can hold.

INCARNATION

You say what I say when I come
doesn't count, as if passion itself
were deceptive. But the root
of passion is to suffer, to allow.

What I want to tell you has nothing
to do with religion, although it feels
sacred: why disallow what bursts through
my mortal lips while we are enveloped

in God's garments? What I utter
at the height of passion—*love*—
is pure, is true, since the flawed
self that I am bears up, is borne away.

CRAZY JANE'S RETURN

Crazy Jane.

Love, when after love we lie
our booty spent, still
your languid fingers toy
at sites of merriment
and still you tarry by my side
though your rod be bent.

You used to scuttle from my bed
at first cock's crow, but now
you linger, yawn instead
and do not go, you do not go
as if you're paying homage to
love's work in field and furrow.

Wild Jack.

I know we both shall rise again
in no mean hour's time
to squeal as do our barnyard kin
(though some deem it a crime)
at holy trough and sacred sty
where the soul roots, sublime.

Many former loves I've known,
many with biblical names,
but none could heal or satisfy
like my Crazy Jane, and none
was more wholly sane.
There was none more holy sane.

Crazy Jane. Oh crimson is your tongue, your speech,
what would the Bishop say?
Never Heaven shall you reach!
where the likes of him do bray.
But Heaven have we found on earth,
down on all fours we pray.

A MATTER OF CHOICE

"Pet or meat?" the braless
woman asks. Under each arm
she cradles a rabbit. This

is a woman who knows how
to kill: "Just lay them
on the ground and lop off

their heads." And how
to give pleasure—her two
thumbs press a vein in each

rabbit's neck. Love,
sometimes when you enter
I bare my teeth. Salt

or sweet? The body decides
since the body must answer
to curious, various hungers.

SUPEREGO SERENADE

Call me vicious rigid but each and every
moment I'm beseiged by want and want and want
you greedy child your plump fist in your mouth
if not for me you'd eat a tree a house gulp
down the very world itself and I know even if
you don't that inside every fat woman is
a fatter one so bless me now and keep your foul
tongue furled inside your mouth I'm here to teach
control to take control and if I sometimes squat
to hold you down I'm no pansy I can take your
rage and dish it out you pig you cow you two bit
whore of Babylon who'd take a crack at every
hunk who's got two balls inside his pants it's up
to me to ladle guilt and blame and don't and don't
to chill you with an image of your mother who's realer
than your mother though she'll be blowzy cheap
and flicking ashes on your cheek you'll know
she's out to kill you with that dagger tucked
inside her stockinged thigh her smartly crayoned
eyebrows cut a jagged pathway to her temple so
reap your just desserts for lusting after taken
men for boys for old old men and as an encore
I'll send in my hit squad to gun you down and for
the mop-up job here comes the crocodile with snapping
jaws you'll wake up with the sweats but fess up
you're relieved to count on me sure as shootin'

FORESHADOWS

Now that I've read
the first draft of your body
I can already tell
the rough places
where there is need
for further exploration.
In due time the clenched
jaws will release a flurry
of words, unearthing
the grave heart
from its vault of ribs.
And since there already is
perfection, the soft mounds
of your fingers, the exquisite
way they interpret my spine,
well then, I can wait
for the rest, the story
your tongue will tell
to the flesh and its slow unfolding.

II

HEROICS

Love calls for a great act
of the imagination. We haul
ourselves out of the brackish past
and again it's by touch
that we know it. Awash

in rapture the human heart
cries for expression, as Aida
and Rhadames hasten by singing
their own demise. Love, our story
has all the ingredients

of high drama—the night sky
ablaze with scuds and you
trapped in a sealed room,
tachycardia. Meanwhile, I drive
my wasted mother to hospitals

which refuse to admit her.
Events conspire. Impossible
to discount the bodies we're heir to.
The day before my mother's lover died
he seized her hair and tried to mount her

from behind. Your father
at eighty-four stripped himself naked
and made obscene gestures.
A ludicrous passion survives. Nobody
tells us this. No consoling arias.

IN YOUR ABSENCE

At first it was your body
I missed, its trappings. This
has not changed. I still feel less
composed without you. But the weight
of your presence I do not miss. Nor
your shambling through my house.
I do not. Miss it, I mean. And here

we thought inertia was the way
of things. How lethargy has
its own momentum. The world is ripe
with calamity. I have never
slept better. Where you are—
the holy land, the hotbed
of disputed claims. Sharing quarters

with your wife. Debilitating meals
at the kitchen table where your heartburn
vanishes. "We're used to it," you say
of war, as though it were weather,
could be sealed out. Even
the keyhole. What we had. I mean
we slept through it all, the fall

and its implications. How spacious
absence is. Restorative. Elsewhere, and long ago,
mammoths neared the land bridge, lifted
one lumbering foot after another,
sustained by arctic ice.

WAR DOESN'T WANT

War doesn't want to be
an arcade game, doesn't
want an enemy a blip
on the radar screen.
It doesn't want a victor
with the best eye-hand
coordination. It wants
the thrill of killing
at intimate range, wants
the torch, the stench
of singed flesh (the skin
tastes best). It hates
dining alone, making death
with strangers. It wants
to know what it's eating.

FEEDING FATE

Once we circled a deserted lot
to avoid a black cat crossing
our path. I indulged you

and didn't laugh. Grateful,
you confessed your other fears,
bargains you strike to clarify

the future—if the light turns
green when you reach the corner
you will reach your destination.

You read accident as billet doux
from the fates, fates you see as jealous
and extravagant. They love love

but exact payment. So yes,
I yell at you over transcontinental
phone lines, call you stupid when

you say you gave your wife lots
of cash. "How did she look?" I ask.
"Fatter than before, but," you add,

"she wore lipstick." I'm amazed
you took in that detail. And me,
how do I see fate? As a wild

woman, swollen with rage, her lips
smeared magenta. But once paid off,
pacified, relinquishing a man

I thought I must have or die,
an aging stubborn man whose faith
in me is more than I can swallow.

*GET**

You entered the rabbinical court, you
who loathe everything that smacks
of the shtetl. Outside, the oppressive
August heat, inside, greedy lungs gasping
for their portion of air. And there
was your wife, her lowered head,
black scarf hiding the hennaed hair.
The rest of her, too, was cloaked
in black. You must have looked
away, flinched at the sight of her
dark red heart. Lawyers, rabbis, wife,
you think I don't know how it was
for you who spit three times
at *yeshiva bochers* scurrying
to prayer? That feeling of choking,
divest, divest. You wanted to peel
the flesh from your bones and start over
fresh. You called me that night
with an urgent plea—*marry me.* As though
I could name you, wrap you in swaddling,
breathe your breath.

*Rabbinical divorce

HUNDESLEBEN

I

Red ones. White ones. Yellow
and green ones. My mother lives
from pill to pill, a pill
for every occasion. For pain.
Sleep. Swelling. Nerves.
For blood. Bowels. *Hundesleben,*
she says. A dog's life. Years ago
it was Chuckles, sweet jellies
at the movies. She wouldn't eat
the green ones. Chuckles. Now
pills. We all know of the black
purse she keeps by her pillow,
her stash. And let her. Even though
they make her wet, cry out in Czech
for her second dead husband. Swaddle
her in pampers, and let her. Once
we counted forty blue pills clustered
on her sheets. Forty. For sorrow.
Forget-me-nots. Toxic bouquet.

II

During our time apart, I committed
unnatural acts: soaped my mother's
bald pubis, shoved her rectum back
into the cavity so near the one
that birthed me. She cursed me
in Czech, called me a lowdown cur
when I denied her Halcion. Seeing you stoop
to lift your battered vinyl suitcase,
no one would have faulted me had I chosen
to ship you back. Not you. Especially
you who would have said the fates delight
in sport. As it was, a pure gladness
rose up in me, and it mattered not at all
that you were not rich, not young, not
savvy. I wanted to take you in, and
be taken. They say it is always
the wrong person. You are not
the wrong person. You are not my mother
though you bring me your failed flesh.
You are not my father whose severity,
he thought, would train me for life. No,
you are a simple man whose first allegiance
is to the parched land of origin turning
evergreen and succulent from *hundesleben,*
this worrying and burying familiar bones.

A SMALL MATTER

To see clearly, first
one must make things complicated.
Which is why the Chinese delight
in labyrinths. And why
they fill their homes
with many children. Only demons
walk a straight path. And I
who lived in solitude so long
it fits me like a slipper. It's not
your barbs that turn me wild,
it's your cursed meddling, reminding me
that you've unpacked. Only God
sees what's behind that nude
reclining on the grass. Perhaps
the one who used the brush.
Small wonder, then, this rage
for order. A lie, but simple.

FAIR RECKONING

The fear is always ending
with a loss—of heart, of hair,
of faculties. That's why I blanch
to hear you've got twenty shekels
in the bank. Who said love
was cheap? Where you come from
they say you squandered all
for love. Likewise, I
was warned not to count
on it to save. That's rich.
We're counting now. Squabbling
is our way of penance. We pay
with pleasure, for it leaves us
beholden, washed clean of debt.

ANNUAL LEAVE

So this is paradise, what paradise
is left to us, earned by the daily
compromise we call work. A breakfast
of mangoes and bacon, fried to perfection,
under a canopy of green. Even the birds
are plump and well-cared for, gold ones
and red ones, obedient children, pecking
at feeders for our delight. I learn
their family name: *fringillidae.* The bass
in the lake are just the right size, not
too big, not too small. We return them
to water to catch again later. It's said
there are snakes here, copperhead
and rattler. They keep out of sight.

And on the second day I ask, lips glistening
with juice, why you always play the black sheep.
Stunned, you strike back: and why do you always
sell yourself cheap? Hard words, hurled
like stones. But didn't we come here fugitives,
shopworn from that other life, its lexicon
of insult and accusation? And isn't this
paradise, a place to set forth from
naked, headstrong, newly entwined?

III

MARVELOUS PURSUITS

1. Overture

A man with two mistresses desires
nothing but peace and tranquility.
He suffers if they should quarrel,
preferring their foreheads unwrinkled.

One day he receives a letter.
The words leap out in bold
typeface: "Cuckold. Cuckold.
More to follow. A neighbor."

He cannot imagine who sent it.
He is obsessed with curiosity.
Which neighbor? Which mistress?
Both mistresses occupy luxury flats.

Both mistresses occupy luxury flats.
Rosa's furnished in bone-white and black,
and Blanca's in all shades of rose.

Rosa and Blanca, like two sisters,
one naughty and hot, quick to anger,
the other so plump and luxuriant.

On her rose settee sits Blanca,
her fingers caressing the keys
of the old Royal typewriter salvaged
from her dear late father's estate.

On her rose settee sits Blanca,
puffing slowly on a cigarillo.
Soon she will put on her poncho,
take an afternoon stroll to the maildrop.

2. Wanton Modulations

The hydrangea, having never been
so purple as this particular year,
cast purple shadows on the cobbled
walk which leads to the front porch
of the house on Pompano Street where
Marta rocks in rhythmic frenzy on slats
that slope and creak from dry rot.

This morning as her husband donned
his starched white shirt, she noted
purple bruises on his shoulder blades
like dark cherries, a message, perhaps,
from either one of his two mistresses,
intended, surely, for her eyes alone.

Jealous? No. She knows to him she is
the one who darns his black silk socks,
takes too lax a hand with kitchen maids
who fail to dust the parlor to perfection.
Let him dwell on youthful breasts, sweet
abundant flesh of those who hold him gaily.

Daily, she bids her spouse a grey farewell,
retires to the rocker, her faded housedress
a once-wild floral print. In undergarments
clean and frilly, she awaits the piano tuner
who comes faithfully at noon when cobbles
on the walkway are their own color.

3. Leitmotif

The piano tuner never comes empty-
handed. Today he brings a tiny tart
of marzipan and raspberries. Perched
sedately on white wicker, he observes
formalities, inquires of the kitchen
maids, their foolish superstitions. Marta
tells in tremulous trill of the wounded
bird who found its way into the larder,
how she herself released it. The piano,
he asks, how has it been behaving?
He plays a slow arpeggio, cocks
an ear for off-key tones. He has
his way with strings, tunes them taut
or slack until each note rings out
according to its true pitch.

4. Seasonal Variation

The piano tuner's mother foretells
the weather. Like a bullfrog,
her arthritic hip senses rain.
Nothing to be done but make a trip
to the doctor who packs her in
medicinal mud imported from Ischia.

Alas, it has come to this: athletic
limbs that danced tarantellas until
daybreak, paddle about in support hose.
Of late she's taken to her rosary, and
a small nip of sherry before vespers.
Her angelic son, runt of the litter,
plays Scarlatti after sunset, balding
pate glowing like an unhatched egg.

5. *Inverted Turn*

With cold-blooded haste, he breaks off
contact with his two mistresses, takes
a chamois cloth and polishes his ox-blood
wingtips, revealing the tiny pin-pricks
like pores on the nose of the shoe, visible
as the blue tattoo of a songbird—a thrush—
on his chest, *Marta Forever* indelible,
inscribed just yesterday on his willing flesh.

Overcome by fever, he must take
to bed, advised by doctors to abstain
from physical exertion which might lower
his body's resistance to fight the strain
of infection. Marta shuffles in slippers,
brings fruit juice, wraps a cool compress
on his wound. If she feels desolate,
trapped, she doesn't show it, nor regret
for her affair, exposed, thus ruined.

His wife, so recently reviled and scorned
by him for newer, more exotic flesh,
becomes by infidelity transformed
into the forbidden erotic object
of his single-minded lust. Her ample
girth once filled him with disgust, the sheer
expanse of it, but viewed in betrayal's
light, is suddenly, irresistibly, dear.

6. Perfect Interval

Rosa, supine, takes the last
rays of sun, the last drops
of oil, Blanca's insistent

palm lifting flesh from bone.
Muscles contract in the cave
of the groin as Rosa turns,

spring lamb on a spit, feels
a quicksilver tongue moisten
her lips. She is used to being

adored, the momentary shadow
her lids absorb, that grip. She
answers with hips, thighs, clenched

toes, finding in a state of tension,
repose. A circular brick wall
permits this feline world of curve

and claw where orchids bloom
multitudinous as weeds, concealed
from marauders, their scorched eyes.

7. Capitulation

And wouldn't we always always slip
into the hot slick coat of it,
spurning food to feel our mouths
go dry with it, stretch out
our arms for this one's lust,
that one's need? We tremble
into being, take in the morning's
striking palette, our lips dusted

with the taste of it. Those sounds
we make, wouldn't they be torn out
by the root, and the wounds
suffered gladly? Anima, animus, we
descend into our evolutionary niche,
wild, demonic, from the bliss of it.

IV

FUN WITH DICK

If to be in life is to make
choices, I'm in it. So says
my private eye whom I'll call
Dick. Dick and I hold this
rendez-vous under potted palms
in a posh hotel. It seems
I've chosen this dime-store novel
plot to play in. Dick states
my best shot: he tracks my husband
to Lulu Labelle who believes
in knots. Or some sweet thing
who's under twenty. Or a guy.
Something he wouldn't want
to get around. And then,
says Dick, you swap. Mexican
standoff.
    You know, yawns Dick,
slouched in a barrel chair, your hubby
called me. To follow you. Such
a big world, I think, and only
one Dick. I trust Dick, more
than lawyers or lovers. I trust
his stale clichés, their sudden ring
of truth, like koans—*fight fire
with fire. It all comes down
to dollars and cents.* I'm in
a fix, I say. No, says Dick,
you're in life. Tonight he'll go

hear Milli Vanilli, teeny bop
rockers. Ensconced between the wife
and daughter, he'll plug his ears
with cotton, whistle Cole Porter
under his breath, a beatific
smile flirting on his lips.

I'M TOLD

I'm told he dates a woman now
who looks like me, but is described
as secretive, a woman who lost
custody of her child because
she beat her. I'm told the girl
is grown now and very beautiful.
You'd think I'd be appalled, or
gloat over his misfortune, but

who am I to judge acts committed
long ago, and from a different
kind of terror, when a woman turns
against her own flesh, once removed.

GIRL TRADES BABY FOR PROM DRESS

—National Enquirer

How smug we feel, we who measure want
in dollops, but imagine her, wanting
to distraction and to be so trapped.
That's how much she wants the blue
chiffon with blue spaghetti straps,
blue rose pinned to the waist and shoes
dyed to match. She wants the gardenia
wristlet, to dance under the bluest moon,
a chance to be elected queen, to neck
in a rented limousine. A horse
isn't hers to give, nor gold,
nor tea. Kings have traded more,
been called heroic. When Anna
abandonned Sergei, wasn't she tragic?
These days I feel my own parched
throat, the bedrock crumbling.
For the fecundity of your embrace
I too would pay, and dearly.

BOLT

For years you've lain flat
on your back, letting life
take you like that, you might
as well be the damask tablecloth
where guests set their plates.
Then one day you shift a little
and the tumbling begins. Nothing

dramatic, though the dramatist
in us all craves a more flashy
event—a lightning bolt and the roof
under which you huddle is fashioned
of tin. Yet always somewhere
in the world it is storming
and always each morsel you down
contains a mineral trace.

UNDERSTUDY

After her second husband died
I thought my mother would
at last define herself, stop
the hard work of being adored.
I was wrong. Within a year
there was another man, colder,
harder, like the diamonds
he used to cut in Bruges. Now
he's 94, reduced to outline,
using the jeweller's loop
to decipher the headlines.
Arthritis has altered my mother's
smile to grimace. To see them
holding hands, their heads
slumped forward as they doze,
is to see two shadow selves
barely tethered to earth.

To move forward into your own
life, first, they say, you must
resolve anger. Meaning, I thought,
to forgive. Some things cannot
be forgiven—Mother banishing me
from Father's deathbed, Father
striking out when I was little.
I only played a bit part
in their dramas. What was I
rehearsing for? For this,

to help my mother hobble
from bed to vanity
where cosmetic jars still line
the counter? She will not,
I know, reach for the magnifying
mirror where once she checked
her expression. With me
beside her, she never could
bear the sight of herself.

EVENSONG

Mother, I watch the tiny o
your mouth forms, barely wide
enough to admit a few grains

of rice. Your elbows tucked
inside your sleeves, you eat
like a bird, and like a bird

your black eyes dart while you chew.
Once you preened at the dinner table,
used implements deftly to remove

the skin from a chicken breast:
"I can't tolerate fat." And Father,
sullen, overweight, accepting

the seconds you heaped on his plate.
After he died you played widow
to perfection, too singleminded

to groom me. Mother, no one
cares to snatch your feed. Soon
the whitethroated sparrow will

claim you for his own. Don't look
at me as though I were the predator,
as though I would peck at your eyes.

LAND OF THE DEAD

It's hard to recognize a soul
in this sea of corpses with no
distinguishing marks. The dead
do not assist us. It would help
if they were stacked in some kind
of order, according to last
initial, or degree of sin.

If I were to call out *Father, Father!*
would he know my voice, his bones
rise en masse, press close
to my still-warm flesh?

Father, I know now we are imperfect
inventions. You never touched me
except in anger. I too have struck
my own children and not repented.
Where does it end, this impulse
to wipe their milky faces clean
of adoration, this refusal
to suffer unblemished trust?

THERE FOR THE GRACE

There for the grace of god there
for the grace there for the window
the housekeeper carelessly left
ajar for the five-year-old boy
who fell from the fiftythird
floor there for the musician his
father and his guitar there for
the mother the mother and there
for the cruise ship out for a cruise
for the cripple whose wheelchair
was pushed and tumbled down into
the Mediterranean there for the blue
of the sea of the starling and finch
and the hyacinth there for the lover
who follows the script who touches
the lips and within the mystére there
for the sweet inhalation of breath
for the arm casually raised to brush
back the hair there for the grace
of god for the grace of earth we go

V

WHAT GOT LEFT BEHIND

EVERY YEAR EXPENSIVE LEATHER LUGGAGE

lined up like soldiers in the hall.
Outside, the taxi beeped its horn
and the old lady rushed downstairs,
her string bag bursting with odds
and ends: cobalt blue Noxema jar,
Kodak instamatic, a whole Polish
salami. Her son raged, "You can't
go like that, a peasant woman off
to market," threw the contents into
a Pan Am travel bag. She is off
to Fleischmann's in the Catskills
for her summer stay, where Olga
waits to blast a brassy welcome
on her trumpet. The string bag,
relic from across the ocean, he
wadded up inside a fist, invisible.
Expanded with necessities, nothing
ever fell through its holes.

AUGUSTA AND OLGA EACH HAS HER OWN

private room, but visit often
late at night, confide deeply
from the heart. Augusta tells
about her son—he wears suits
from Knize, shirts from Sulka,
pats his face with Aqua Velva, yet
he still picks his nose in public.
She hates his socks, black silk
held up by garters. She knows she
should be grateful, after all it
can't be easy. Olga tells about
her grandson, Stephen Miadovnic.
They make a match with Augusta's
granddaughter. Wouldn't it be
wonderful? Each sighs contentedly,
imagining the wedding. All around
the Catskills, pine trees shiver
in the breeze, scent the midnight air.

MR. KAHN KNOWS HE'S NOT MUCH

to look at, yet he has one
virtue: he has lived long.
Thus he counts his blessings.
He has his heart set on Olga,
loves it when her lips purse
around the trumpet's mouthpiece.
He hoards his silver dollars
and his sourballs. Tomorrow
he'll present her with a purple
treat, watch as it dissolves
inside her cheek. Gently
he strokes his shrivelled sex,
astounded by its slow ascent.
Outside cumulus clouds drift
by, stars emerge and multiply.

OLGA IN THE CATSKILLS

is careful to keep her rosy skin
protected from the sun, deals
canasta till all hours, flashes
photos of her grandson. Sometimes
she plays a rousing trumpet solo,
stockings wrinkled at her ankles.
Mr. Kahn admires her, tells about
the three Jews interrogated by
Saint Peter as to their fidelity.
Olga's laugh comes from her belly.
One day she catches him cheating
at cards, refuses the silver dollar
he offers as apology. "Now I can't
ever trust you as my partner."

MRS. FLEISCHMANN HAS A SORROW

she carries with her constantly.
Her son, born with flaxen hair
and skin as pale as bone china,
has married a Negress. Not
café au lait, but black, coal
black, black as the ace of spades.
She's received a letter—they wish
to come to help her out this summer.
Marietta is with child, and mountain
air will do her good. The pity is
that Mrs. Fleischmann likes Marietta,
thinks her fine and able-bodied, looks
forward to her company. She wonders
what went wrong in Paris, what her
guests will think or say, if she'll
hear them romping in the upstairs
bedroom where she's changed the sheets.
She gazes out the kitchen window, sees
her daughter Charlotte stop the swing,
pick a scab, begin to braid her hair.

THE TRUTH IS EVERYONE DELIGHTS

in Marietta. The ladies bring her
soda crackers, inquire how she spent
the night, tell her tales of labor.
The gentlemen are courteous, attempt
to bring their plates in, but she
just laughs and shoos them back.
Charlotte follows her around,
seems to like to touch her skin.
The truth is, it's her son who
worries Mrs. Fleischmann, the way
he lounges on the porch, a deadened
flat look in his eyes, not bothering
to read the mystery thrillers he
used to love so much. Marietta
says he just needs time, that
something broken must be mended,
so Mrs. Fleischmann doesn't pry,
lets the sun-soaked days slip by.

AUGUSTA CANNOT BEAR TO SEE

Mr. Kahn so...reduced. There's no
other word for it. His trousers
flap, formless in the summer breeze.
His face assumes a hang-dog look
which is most unbecoming. Olga
treats him as her lackey, demands
he fetch her book, her shawl. He
complies, no longer tells his jokes.
Olga's not so perfect, bolting down
her fricassee as though there's no
tomorrow. She'll speak to Olga,
explain how one should be in life:
merciful, make allowances.

A MOST MARVELOUS SIGHT

greets Mr. Kahn one evening, when,
restless and insomniac, he strolls
the hotel grounds. Silhouetted
against the gauzy curtains, Augusta
gives herself a sponge bath. Her hair
released from its bun streams down
onto her shoulders. Her pendulous
breasts bob and sway, bob and sway,
he has an urge to nuzzle. She lifts
one fleshy arm, with the other, laves
the hollow. Never has he seen such
breasts, his Irma's small and bitter.
There, firmly planted under moonlight
with his head flung back, fondling
himself, Mr. Kahn is a happy man.

EVEN MRS. FLEISCHMANN NOTES THE CHANGE

in Olga. She's grown querulous
about the food, and hardly talks
to Augusta, who now walks hand
in hand with Mr. Kahn, oblivious
as a June bride. Olga snaps
when she plays cards, and plays
only to win. Summer is drawing
to a close—the evenings hold
a definite chill. Her thoughts
turn to insulation; soon she'll ask
her son to check the storm windows.

OLGA IS APPALLED

by the lengths some women go
to trap a man. Take Augusta,
she'll laugh at anything that
spindly Mr. Kahn says, covers
her mouth with one plump palm
as though he's too, too naughty.
Olga can't be sure, but thinks
they are "doing it." Augusta
has no time for old friends, and
time is one thing none of them
can squander. Who knows where
they'll be next summer, who'll
break a hip, contract disease,
whose blood will fail to clot?
Olga finds her thoughts oppressive,
no longer lingers after meals, takes
her tea out on the porch. There she
finds young Mr. Fleischmann. Sometimes
they speak French for fun, do the Sunday
crossword puzzle, name exotic beasts.

OLGA'S HAD A THOUGHT OR TWO

about young Mr. Fleischmann, imagined
lying next to him, cradled in his lanky
arms. But she, for one, can separate
fantasy from real life. He's composed
a duet for them for the annual talent
show. A gifted pianist, he's starred
in numerous recitals. They rehearse
a kind of blues for birds returning
to their nests, bees who've ceased
their buzzing, until the dinner bell
is rung, guests are summoned to the fold
and fireflies swarm in radiant dance.

IT AMUSES MR. FLEISHMANN TO KNOW

the shotgun's in his trunk, swaddled
in a blanket. This morning loading
groceries, it was there. And now, arm
draped round Marietta at the drive-in,
a secret presence, like another woman.
It makes him feel less trapped, as though
he has some options. "Imitative. Vulgar."
That's what *Le Monde* printed. For days
he couldn't drag his body out of bed.
He idly hums some notes from the duet,
makes a shift in the progression, decides
to end the piece in a minor key.

CHARLOTTE INVITES MR. KAHN TO VIEW

her rock collection, labelled
with adhesive tape. A chunk
of celestite sparkles bluely
in her hand like underwater
stars. "I thought I'd want
the amethyst, but I was wrong,"
says Mr. Kahn. Celestite, color
of Augusta's eyes, and the cotton
dress he loves with scattered
periwinkle. He digs inside his
pocket, extracts a silver dollar.
"I think one is fair, don't you?"
Charlotte squints into her magnifying
glass, frowns and says, "Maybe two."
"Throw in the rhodochrosite and I'll
make it two." He'll keep the pale
pink rock, light as strawberry mousse.
The celestite he'll gift wrap, tie
it with a golden string. The index
card in schoolgirl script reads:
Bought at Chittenango Falls.
Crystal system: orthorhombic.

AUGUSTA SITS ON THE EDGE

of her straightened bed, inspects
the aging portrait. Perhaps
she presumes. Why should he care
how she used to look? Horrid hat,
she sees that now. She never was
a beauty. But skin without a wrinkle
and eyes clear and hopeful. Cameo
brooch, heirloom from her mother,
where is it now? And Schatzi, beloved
dog, long gone. This daguerreotype,
its edges furled, the sepia intensified,
perhaps he'll treasure it, remember how
he made her feel: virginal and ripe.

EARLY MORNING LIGHT FINDS MR. KAHN

with pen in hand, carefully composing
an ode for Augusta. He's already wrapped
the celestite. Now nestled in a velvet
box, eleven glowing sourballs, one for
every month of separation. He sees her
rocking in a chair, outside a gusty wind,
but within her mouth, snug and moist, his
candy melts. He writes: "For you I'd buy
crown jewels. Hope these will do. Taste
and remember my undying love for you."

CHARLOTTE TRAILS HER FINGERS

along the window sill, taps her foot
in time to a hit melody. She's bored.
The guests have gone. The taxi bumped
up the hill this morning, and then
bumped down. Upstairs, she investigates
the rooms. Not much this year: a playing
card, the Jack of hearts. On its flip side,
a fancy woman dressed in miles of cloth,
puppy in her lap, and in the corner, an...
umbrella? An empty jar of sapphire blue.
She'll use it to store her foreign coins.
Some scraps of paper, it's hard to read
the writing—"For you I'd climb" crossed
out, "I'd give, I'd buy." Some cotton
balls. They'll come in handy next time
she applies mercurochrome. Each object
catalogued accordingly. She lies down
on her bed, finds her silver dollars,
bites them till they hurt her teeth,
arranges each over each closed eye.

VI

THEIR SPOUSES COULD PLAY

Their spouses could play canasta at the cemetery
is how the joke went the summer four members
of the refugee circle died. One of the players

would be my mother, chic in her widow's weeds,
slapping the red threes down. And how the dead
would kibitz, tally the pebbles the living left

to mark their visits. Herr Glauber came courting
before the first frost. The night of his wife's
burial, the phone rang and my mother and I answered

at once. I heard them discussing blueprints and the bidet
for the new master bathroom, and that I must be sent to live
with his pregnant daughter. And so it came to pass—

my senior year at the house on Juno Street, witness
to my stepsister's swelling belly, and I was not
unhappy there, the small seed of grief tucked

in my gullet. It flourished over the years
and I tend it even now, as my mother lies dying
in the remodeled bedroom painted dark teal, scenes

of the tropics splashed on the walls. I have never
spoken of this, the exile from the house my father
built, my mother drugged on a rented hospital bed

and the vines that grip my cold and capable heart.

ALARUMS AND EXCURSIONS

Distressed and distraught over a broken heart
a woman decides she needs protection. Enter
Francis Drake from Be Secure. What I'm selling
he says, is peace of mind. The system he touts
is used by the navy and NASA alike. Something
like killing a fly with a bulldozer, she says,
since she's mainly concerned about her back
door. Like his namesake, Drake's been around,
has even defeated armadas, pirates greedy
for foreign gold. He calmly recites drastic
scenarios—taken hostage while carting groceries,
stungun for the dog, the fire that travels
90 feet per second. Or, he says, you could
fall downstairs and break both legs. How
do I know, she asks, you're not a serial killer
or a scam artist? Adele, he says (her given
name), you don't. She notes his fingernails
trimmed and clean, and his eyes a benevolent
baby blue. The power surge, he adds, is a sweet
touch, as she imagines his and how it would feel
hitched to a man who knows the tricks of breaking
and entering and how to secure all she holds
dear from those who would trespass against her.

THE PSYCHIC LOCATES THE VICTIM

With these hands I touch her blouse her scarf her
lacy underwear   hold them
to my nose   sniff
the scent of musk and talc
let it lead me where
she's bound and gagged
girdled   no girders
bound and gagged somewhere
where vertical shadows slash
cement   floor   factory
something stiffens   deafens
periodically   on schedule   alarms
ring in hours   subways   someplace
industrial   he checks the
knots   pendant   gold fob   watch
dangling from a breast a   vest
pocket   picket   fence unpainted
slanted   derelicts
lying by a gate   a grate   steam
rising from their heads like
halos my god I'm homing in closing

in my god she wants I want
to shit   the fear   feral   evil
will he   will he let me   live
or slice me open   melon
will my blood trickle   spurt   flood
obsessed with blood   with visions
of my blood   I want
to keep it   in   my blood
my shit his thin neat as a pin

vertical   ramrod hard he
holds the card the ace of hearts
he weaves below my chin
his jaw   glistening   beads
of sweat like the steel
blade the knife he   toys with
cleans his fingernails   intent
he's forgotten what   me
who knows   maybe my shit would turn
him on   is this what wolves feel
when they bare their throats   fling
their heads back   fact
I'd throw my head back   part
my lips my limbs   take him
in   obedient   grateful   praise him
for sparing me my god he
plunges in   the ropes
around my wrists grind through
my skin I take him   in
he grinds his hips   I   mine   thrusts
his pelvic bone on bone   mine   I
cry out   praise him   praise
I'm still alive   he spared
my face my chin   I take him
in   and in   I she she's
please let me   out   the smell
of musk and talc   smothers   mother
breathless can't catch   my breath
so tired now please let me
sleep   I see it now she's
lying there her hands
still tied   alive
used up   used lot
on 27th Street.

MIAMI VICE

In a stylized dance, the villain
traces his stiletto from breasts
to throat to web of flesh behind

her ear. "Mine," he breathes,
"every inch of you." Her liquid
eyes lock with his and when a drop

of blood appears, signifying
rupture, we eat it up, we love it,
we love him, dazzling in his peach

linen suit, spitcurl glistening
on his forehead. We love his black
Maserati, his world of pastel sky

and scanty dialogue. He comes to us
in sleep as though on assignation,
cigarette dangling from his lips,

our own mouths slack and unrepentant.
He brings a bellyful of unsettled
scores, makes our muscles gear up

for flight. We never knew we could
feel this way, we never knew our blood
to be so brutally red, so gorgeous.

TALE FROM THE NORTHWOODS

Charles Van Riper, pioneer
in the field of speech-language pathology

Once by drowning, once by hanging, once
by inhaling carbon monoxide, three
attempts before I was thirty, all

failures. All because of a tangled
tongue. And a father who beat me
for the filthy habit, calling it akin

to masturbation. The summer I pretended
to be deaf and dumb. The quack cures,
cruel experiments, needles in the mouth,

psychotropic drugs. Singing. The ordeal
of going to the corner store for Dill's
pipe cleaner. Of walking out without.

Aspirations zero, until I hitched a ride
with a farmer who said don't labor so hard,
just stutter better, let the syllables

leak out. I might have been St. Paul
on the road to Damascus. Soon a man
appeared in spotless attire making odious

sounds. Took him to a calving. Up
to his shoulder in the bloody muck.
Made him extract the afterbirth, dump it

in a bucket. The man turned sloppier,
the stutter, cleaner. Every case
a metaphor to deconstruct, unleashing

the unsayable, the choked back flood.

WHAT IS SERVED

There must be an appetite for grief.
How else explain this gnawing on memory
until its bones are clean? The soul
barks and barks for its ration
of remembered things—the smell
of tangerines, undergarments stripped
like husks, exposing flesh as willful
and imperative. And how surrender
nourishes deliverance.

Now we are
swollen with absence and abstinence
tucks in our sheets. In our mouths
the aftertaste of satiety. Narrow
the bed we lie in, ample, our need.

STORMY PETREL

for Cully Gage

Before she must leave for the long
flight East, the old man insists
they descend the cellar steps, past
hand-hewn beams, past the massive
tree trunk he calls Atlas. Here
in the wine and fruit room, ten years
after his wife's death, are mason jars
filled with her pickles and peaches.
Beyond, in a closet, rests the skull
of an Indian girl. He calls her Nokomis
and tells how she died from a blow
to the right ear. He shows the woman
the hole. He discovered the skull
in the tangled roots of an upended oak.
No other bones. Wolves carried them off
says the man, no stranger to woods or habits
of wolves. But the skull wouldn't fit
in their great jaws so he took it home
to this dark cellar in Kalamazoo, childhood
name for nowhere. And places it now
in the woman's palm. She runs her fingertips
over the burnished bone. It is not cold.

*

I've been thinking, he said, only
a few hours later, during a meal
of venison sausage and warmed-over
pasty, a loggerman's breakfast, what

kind of bird you are (for you are
a bird) and I've finally come up
with the answer: stormy petrel. You dive

again and again into the salty
waves but always emerge triumphant.
The old man smiles, satisfied.
The woman asks if she should change
her feathers. He laughs and says
she couldn't even if she wanted to
and why would she want to.

*

Arriving home, she heads straight
for the Audubon, learns her habitat

is the open sea. Her diet consists
of small shrimp and plankton. She seldom

follows boats. She is jet black,
has a white rump, forked tail, erratic

flight. She resorts to rocky islands
to breed under cover of darkness, filling

the air with nocturnal trills. She lays
one white egg. When ruffled she emits

a musky odor. And this: she contains
so much oil that even after death

if a wick is lit in her throat
it will burn for months. There are times

she has yearned for an ordinary life,
hearth, husband, balanced meals prepared

on schedule. Rues the day she was born
to perpetual flight, and the burning.